MATH and ME

Math at the Game

by Joanne Mattern

LOOK! BOOKS™

Red Chair Press Egremont, Massachusetts

Look! Books are produced and published by Red Chair Press:
Red Chair Press LLC PO Box 333 South Egremont, MA 01258-0333

FREE Educator Guides at www.redchairpress.com/free-resources

Publisher's Cataloging-In-Publication Data
Names: Mattern, Joanne, 1963- author.
Title: Math at the game / by Joanne Mattern.

Description: Egremont, Massachusetts : Red Chair Press, [2022] | Series: LOOK! books : Math and Me | Interest age level: 005-008. | Includes index and suggested resources for further reading. | Summary: "Math isn't just something you learn in school. It's an important part of the world around you. Young readers will learn about counting, adding, and greater than/less than/equal to while enjoying a day at a baseball game"--Provided by publisher.

Identifiers: ISBN 9781643711287 (hardcover) | ISBN 9781643711348 (paperback) | ISBN 9781643711409 (ePDF) | ISBN 9781643711461 (ePub 3 S&L) | ISBN 9781643711522 (ePub 3 TR) | ISBN 9781643711584 (Kindle)

Subjects: LCSH: Mathematics--Juvenile literature. | Sports--Mathematics--Juvenile literature. | Baseball--Juvenile literature. | CYAC: Mathematics. | Sports--Mathematics. | Baseball.

Classification: LCC QA40.5 .M381 2022 (print) | LCC QA40.5 (ebook) | DDC 510--dc23

Library of Congress Control Number: 2021945359

Photo credits: Cover, p. 5, 10, 14, 22: iStock; p. 3, 4, 7, 8, 9, 13, 14–21: Shutterstock; p. 1, 23: Jennifer Walters; p 12: Patricia Walters; p. 24: Angel Santee

Illustration (p. 11): Joe LeMonnier

Printed in United States of America
0422 1P CGF22

Table of Contents

Let's Go to a Game

It's fun to go to a baseball game! You can use math when you watch a game. Let's see the ways you can use math at a baseball game.

MATH FACT!

Keeping score is a great way to use math at a game! Keeping score uses counting and adding.

Buying Tickets

You have to pay for your tickets at the game. A grown-up's ticket costs ten dollars. A child's ticket costs five dollars. Which ticket costs more?

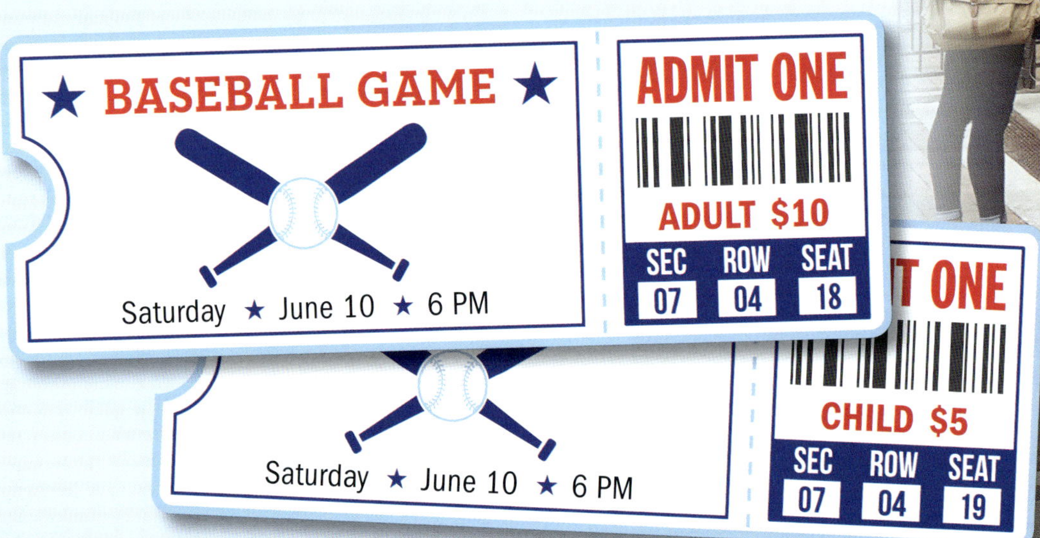

STUBHUB
WRIGLEY FIELD
2019 SCHEDULE
BRYANT
17

Ten is greater than five. Ten dollars is more than five dollars. An adult ticket costs more than a child's ticket.

MATH FACT!

You can write that ten is greater than five like this: 10 > 5.

You can write that five is less than ten like this: 5 < 10.

Team Players

The players run onto the field. How many are there? Let's count. 1, 2, 3, 4, 5, 6, 7, 8, 9. There are nine players on a baseball team.

1. Left fielder
2. Center fielder
3. Right fielder
4. Third baseman
5. Shortstop
6. Second Baseman
7. First Baseman
8. Pitcher
9. Catcher

1
2
3
4
5
6
7
8
9

Play Ball!

The first player comes up to bat. He swings. He misses. The umpire yells, “Strike one!”

The batter swings and misses two more times. One strike plus two strikes **equals** three strikes. He's out!

MATH FACT!

You can write the strike count like this: 1 + 2 = 3.

Keeping Score

A baseball game has nine **innings**. It's the top of the ninth inning. The home team has five runs. The visiting team has three runs. Which team is winning?

Five is greater than three. It can be written like this:

$5 > 3$

The home team is winning.

The visiting team **scores** two more runs! Now the score is tied. Both teams have five runs.

MATH FACT!

Five and five are **equal** numbers. They are the same.

5 = 5

The Winning Team

The score is tied at 5–5. Your favorite player comes to bat. He hits a big home run! Now the score is 6–5. Six is greater than five.

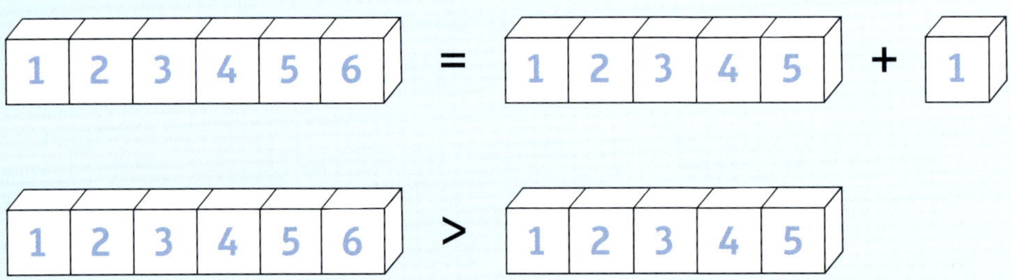

Dodgers
LA
31

The home team won! Did you have fun using math at the game?

Words to Know

equal: meaning things that are the same

innings: parts of a baseball game; a baseball game usually has 9 innings

scores: earns a point in a game

Learn More at the Library

Check out these books to learn more.

Greathouse, Lisa. *Count Me In: Soccer Tournament.* TCM Publishing, 2012.

Levit, Joe. *Let's Explore Math (Bumba).* Lerner Publications, 2019.

Overdeck, Laura. *Bedtime Math Series.* Feiwel and Friends, 2013.

Salzmann, Mary Elizabeth. *Baseball by the Numbers.* Abdo, 2010.

Index

About the Author

Joanne Mattern is the author of many books for children. She loves writing about sports, animals, and interesting people. Mattern lives in New York State with her family.